Information Systems Engineering Library

SSADM and Application Packages

LONDON: HMSO

CCTA
*The Government Centre for
Information Systems*

Acknowledgements The assistance of Phil Lomax of Model Systems, under contract to CCTA, is gratefully acknowledged.

For further information regarding CCTA products please contact:

CCTA Library
Rosebery Court
St Andrews Business Park
Norwich
NR7 0HS
0603 704704

Foreword

The Information Systems Engineering Library provides guidance on managing and carrying out Information Systems Engineering activities. In the IS lifecycle, Information Systems Engineering takes place once the IS strategy has been defined. It is concerned with the development and ongoing improvement of information systems up to the operational stage, and their maintenance whilst in operational use.

The Information Systems Engineering Library complements other CCTA products, in particular the project management method, PRINCE, and the systems analysis and design method, SSADM.

Volumes in the Information Systems Engineering Library are of interest to varying levels of staff from IS directors to IS providers, helping them to improve the quality and productivity of their IS development work. Some volumes in this library should also be of interest to business managers, IS users and those involved in market testing, whose business operations depend on having effective IS support by means of Information Systems Engineering activities.

The Information Systems Engineering Library also complements other related CCTA publications particularly the Programme and Project Management Library, the Information Management Library for data management issues, the IT Infrastructure Library for operational issues and the IS Planning Subject Guides for strategic issues.

CCTA welcomes customer views on Information Systems Engineering Library publications. Please send your comments to:

Information Systems Engineering Group
Rosebery Court
St Andrews Business Park
NORWICH
NR7 0HS

Contents

1   Introduction

1.1 Purpose of this volume	The purpose of this ISE Library volume is to explain the relevance and application of SSADM when the possibility of using application packages, for all or part of the solution, is identified early in the analysis process.

This volume is not concerned with:

- procedures for evaluating a package against information system requirements

- how to select from competing packages (see Appraisal and Evaluation Library volume *Overview and Procedures*)

- how to procure a package (see ISE Library volume *SSADM and Information Systems Procurement).*

1.2   Who should read this volume

The primary intended audience for this volume is SSADM practitioners conducting SSADM Feasibility or Full Studies when the probability of using one or more application packages to meet the requirements has been identified.

SSADM practitioners with the following roles may have an interest in this volume:

- Team Leaders

- Requirements Analysis Data Modellers

- Requirements Analysis Process Modellers

- Requirements Analysis Product Developers

- Senior Requirements Analysts.

The volume will also be useful for Software Development Managers, IS Project Managers and Stage Managers who have some responsibility for SSADM-based IS projects that might deliver application package software as all or part of the solution to business requirements. The volume will give these readers an understanding of how the analysis and design activities are changed as a result of using an application package, and the impact on the lifecycle decision points. Software Development Managers and IS Project Managers within the intelligent customer function in sites where the IS department is contracted out, may also find the volume useful.

1.3	**Structure of this volume**	Chapter 2 provides an overview of this volume.

Chapter 3 describes the different ways of using a package and identifies the key points in the development lifecycle where package options may need to be considered.

Chapter 4 describes how the planned use of a package affects the conduct of a Feasibility Study or Full Study, and in particular how it influences the scope and depth of requirements analysis and specification activities.

Chapter 5 describes how the design should be completed for those systems that need to modify or extend the package software.

Whereas SSADM practitioners will need to study the whole volume, it is expected that many other readers will find it sufficient to read only the first three chapters.

1.4   Assumed knowledge   It is assumed that readers are familiar with SSADM, at least at the level of the Certificate of Proficiency examination, and with the 3-schema specification architecture (although a summary is included at Annex B).

2 Overview

2.1 Definition of a package

This volume is concerned with the use of application software packages, shortened elsewhere to application packages or just packages.

An application software package can be defined as 'a computer system comprising programs that claim to offer a solution to a defined business problem'. This definition is too broad for the purposes of this volume, which is only concerned with packages of the type that contain some pre-defined functionality to support a particular business application (eg accounting; vehicle fleet management), or a particular market sector (eg hotel management, health care).

This volume is not concerned with the use of packages that require the data and functionality to be defined by the purchaser. Personal computer databases are excluded from consideration as they provide facilities to build an application specific system rapidly, rather than using a pre-defined but generic application. Office system software such as word-processors and spreadsheets are also excluded. Although such software products are clearly a packaged solution to a defined business problem, the definition of the data and processing is the responsibility of the user.

This volume uses the following definition of an application package:

'An application package is a commercially available set of one or more computer programs that contains pre-defined data and functionality, designed to provide a complete business application.'

2.2 Advantages gained from the use of a package

There are many advantages to be gained from the use of a package, these are grouped under the headings of:

- cost

- time

- skill

- quality.

2.2.1	Cost	When compared to the development of a bespoke system an application package is likely to cost less in several areas:

- analysis and design of the user requirements

- obtaining the software (buying the package as opposed to writing the code for a bespoke system)

- maintaining the system.

2.2.2	Time	The desire to use a package may be motivated by the need to install something quickly, with cost a secondary consideration.

2.2.3	Skills	An organization may lack the skills to design, develop and implement a bespoke system. Acquiring a package is one alternative to contracting-out the development.

2.2.4	Quality	Packages that have a substantial customer base are likely to be robust and well proven. This can be regarded as a guarantee of quality.

2.3	**Disadvantages of package use**	The use of a package may also have some disadvantages. These are discussed under the headings of:

- quality

- appositeness

- system integration

- control

- competitive advantage

- operational cost.

2.3.1	Quality	There may be some fitness for purpose criteria that are not satisfied, for example, the provision of certain data items or functions.

2.3.2   Appositeness

There may be requirements specific to the organization that are not satisfied by any available package. The decision to use a package almost always implies an acceptance that some requirements will be restricted or modified to align more closely with the solutions available in the package. The business requirements, however, must be properly defined before a management decision is made that will modify or restrict requirements.

2.3.3   System integration

If the system to be implemented using a package is part of a larger group of applications, it may be difficult to integrate the package with other applications (bespoke applications or other packages). It is, however, true that a growing number of application packages now have extensive input/output facilities that meet Open System requirements which makes integration with other software easier.

2.3.4   Control

The use of a package may reduce an organization's control over the system. At the extreme it may not be possible to change the information system when the business requirements change.

2.3.5   Competitive advantage

The adoption of a package that is widely used by other organizations in the same market sector will limit the ability to obtain a competitive advantage from the information system.

2.3.6   Operational costs

Savings in development costs may be exchanged for higher operational costs. These may be in the form of software licence fees, charges for software upgrades or charges for modifications to meet changing requirements.

2.4   Advantages / Disadvantages conclusion

Although the disadvantages may seem more numerous, the use of package software attacks the problem of the availability of development resources which is the most serious challenge facing many organizations in respect of IS development. However, the use of a package is only one of several possible approaches that could deliver the benefits described. For example, the need to implement something quickly could be addressed by an incremental delivery approach. A reduction in cost could be achieved by reducing the scope of the functionality. Deciding when to use a package is discussed in section 3.4.

2.5 Relevance of SSADM	The use of SSADM is appropriate when a project is intending to provide a package solution.

There are those who argue, incorrectly, that little benefit is gained from producing an SSADM Requirements Specification when a package solution is to be used. Whilst it is true that the scope of the Requirements Specification is significantly reduced when using packages in certain ways, the use of SSADM is still necessary, if not, costly mistakes made by choosing the wrong package could occur.

The SSADM analysis techniques are concerned with discovering facts about the system. Discovery techniques are still essential for systems using a package. The requirements still need to be identified for comparison with available packages to establish the best-fit.

The SSADM design techniques are concerned with how the system is to be constructed from individual components; requiring the practitioner to use their judgement and make decisions based on facts and rules of thumb. If the package software is to be extended by additional bespoke elements how are the bespoke elements be to designed?

2.6 Scope of this volume

This volume provides advice on the use of SSADM when the adoption of an application package is intended, including:

- how to modify the structure of SSADM

- how to undertake requirements analysis up to the point that a decision is made to use a package

- what has to be done as a result of a package not meeting all the requirements (ie, treat the package as an already existing system to which interfaces are to be built)

- what has to be done to modify the requirements when the package can provide the functionality but not in the way originally specified

- what has to be produced using SSADM to enable the package appraisal and evaluation to proceed.

This volume does not provide:

- an alternative SSADM structural model for use when a package implementation is planned

- procedural guidance on how to evaluate an application package

- guidance on how to procure a package (see ISE Library volume *SSADM and Information Systems Procurement*).

3 Deciding to use a package

This chapter is concerned with describing the different approaches to implementing a system using a package, and identifying the key points in the development lifecycle where options to use a package may be exercised.

3.1	**Implications of the 3-schema specification architecture**

The 3-schema specification architecture used by SSADM provides a framework for understanding the opportunities for using an application package. Readers who are unfamiliar with the 3-schema specification architecture should refer to Annex B, in particular for an overview of the concepts of Internal Design, Conceptual Model and External Design.

If an application package is adopted, manipulation of the Internal Design to improve performance of an individual application is not usually possible. In most cases the source code is not available. Even if it is possible to change the database structure and Process Data Interface (PDI), the costs of regression testing are likely to reduce significantly the savings expected from the package approach.

The Conceptual Model represents the kernel of the package. There may be limited provision in the package for extending the conceptual data model (for example via user definable data items), but changing the conceptual behaviour model is difficult and should be avoided if possible.

The most common changes to a package are to the External Design. Many packages provide facilities for extending functions and manipulating the input and output data, without changing the content and organization of the stored data or the way it is processed (ie, no changes to the Conceptual Model). For example, the ability to aggregate and summarize data in different ways, change data item names and define additional semantic validation, can be used to make the software appear to match the requirements more closely, without having to attempt tricky changes to the Conceptual Model.

The 3-schema specification architecture is useful in three ways:

- it is a framework for customization of SSADM, (see Chapter 5 for further details)

- it is a framework for requirements limitation, eg performance targets that would require Internal Design changes are not accepted (Chapter 4)

- it is a framework for classifying target systems to establish the suitability of particular development approaches including the use of a package (section 3.2 to 3.4. For fuller coverage please refer to the ISE Library volume *Accelerated SSADM*.)

3.2 Development approaches

The broad development approach of implementing a system using an application package can be sub-divided into:

- package-constrained

- package-based

- using a package as a prototype.

These approaches are not mutually exclusive. In particular it is likely that many projects which start out as package-constrained will end as package-based projects.

3.2.1 Package-constrained

A package-constrained development is where the system developed for users is constrained to only those services that can be provided by application packages, with no bespoke work (other than the possible use of a query processor and report generator on the package file or database).

This development approach can only be used when there is an acceptance on the part of users that substantial compromises over the requirements will, if necessary, need to be made. The package may not do the whole job but the expectation is that it will satisfy those aspects of the system that the users regard as mandatory.

The requirements definition activities for package-constrained systems are limited to the completion of the products necessary to evaluate and compare available packages. To decide on a package-constrained approach too early, before requirements have been defined, can be a very high-risk strategy.

3.2.2   Package-based

A package-based development is where the major part of the system is to be provided using application packages, but some of the user facilities will be provided by developing bespoke programs, and maintaining data outside the package files.

This development approach can be used when a significant proportion of the functionality (typically more than 60%) can be satisfied by the core package. Most of the remainder should be capable of being provided by extending or modifying the package. However, as with the package-constrained approach, there are likely to be less important aspects of the requirement that remain unsupported by the computer-based part of the information system.

Some of the tactics that may be used to extend the requirements coverage of the package without recourse to changing its source code are:

- the use of user extensibility options within the package

- manipulating the input and output data of the package

- the addition of purpose-written enquiries

- the manipulation of files extracted from the database

- the development of purpose-written sub-systems that interface to the package.

Many packages have been developed using a proprietary application generator and can therefore be easily customized (providing a licence for the application generator has been obtained) for specific implementations.

For example:

- screens and menus can be customized to give the appearance of a package that is more pertinent without changing the underlying processing

- the help text can be modified to provide a translation between user specific terminology and procedures and the facilities of the package

- names of data items can be changed in the data dictionary to make them more pertinent (eg changing staff 'grade' to 'rank')

- the report-writer can be used to develop additional reports

- end users can formulate their own enquiries using the built-in query facilities

- data can be exported to decision support tools such as spreadsheets for further manipulation and processing.

The requirements definition activities will be more extensive than for package-constrained projects. However, a full Requirements Specification document is not necessary in every case. The extensions to the package may be limited to additional purpose-written enquiries or features available via user extensibility options in the package.

A full Requirements Specification, including Entity Life Histories and Effect Correspondence Diagrams, is only necessary if some system events are not represented in the package and their bespoke coding needs to be integrated with the package-behaviour model.

3.2.3 Using a package as a prototype

An application package may also be used to implement an initial version of a system quickly. The initial version may be extended or modified in subsequent releases (incremental delivery) or may be entirely replaced by a bespoke system or one based on a different package (evolutionary development).

**3.3   Requirements
Definition**

The use of a package will have three major effects on requirements definition (see Figure 1):

- the modification of the requirements to align more closely with the available solutions. This is a high-risk option if done before the requirements are known as the system when completed may be unacceptable to the users

- the modification of the package to satisfy requirements not met by the base product

- enclosing the package in additional processing to satisfy requirements not met by the base product.

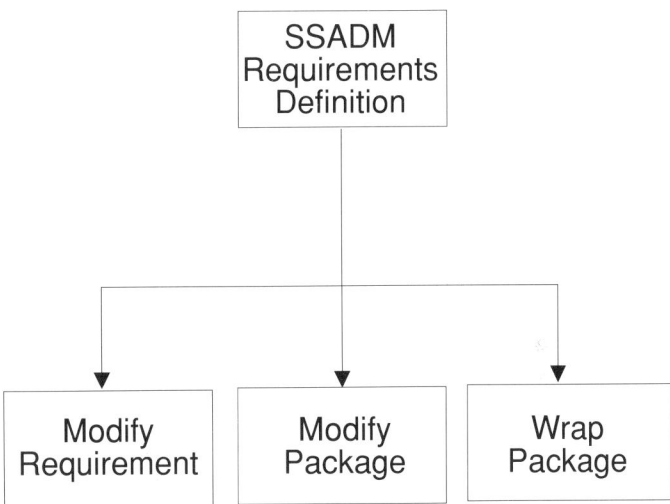

Figure 1: The Impact of Using a Package

A package-constrained approach is only concerned with the modification of the requirements. Some requirements may be satisfied by the package, but not in the ideal manner. Others may not be supported at all. However, a complete and cost-effective solution may still be available by implementing the unsupported requirements in changed working methods, rather than as computer software.

Package-based approaches may include either modifications or extensions to the package. If a choice exists the latter is to be preferred, as it is likely to be less complex to implement and particularly to maintain.

3.4   When to use a package

'Reuse it before you buy it; buy it before you build it'.

The possibility of adopting a package should always be considered as one of the options evaluated in a Feasibility Study, and also in Business System Options if possible.

The main factor governing the availability of software packages for the kind of application under consideration is whether there exists a common requirement and a market. The common requirement may be for business software such as a general-ledger package which has almost universal application across market sectors, or it may be in a vertical market such as patient-administration packages in the health service.

The existence of a common requirement does not necessarily imply high volume sales. Some companies which market financial services packages have a sales target of only one licence per year, but the package costs over £250,000. On the other hand complex general ledger packages for corporate use can be purchased for under £20,000 because of higher sales volumes.

In addition to being related to potential sales revenue, the availability of packages is also linked to their development cost. A record-keeping system does not include a behaviour model and is therefore easier and less costly to build and change. However, a bespoke system without a behaviour model is also easier to build and change, typically using application generator facilities. The cost of the package must be compared against the cost of the bespoke alternative.

The likelihood of suitable packages being available is also influenced by the type of system under consideration. The following types of system represent a subset of system types where both the use of SSADM is appropriate, and package solutions are likely to be available:

- real-time systems

- administrative data processing systems

- record keeping systems.

3.4.1 Real-Time systems

The event sequences in the real-world behaviour model are directly coupled to corresponding event sequences in the information system behaviour model. The sequence of real-world events is absolutely constrained by the information system, for example an airline reservation system will not accept a check-in event until the booking event has been recorded and the seat availability confirmed. Package solutions can be used but they must include the required behaviour model. Such systems tend to be business critical.

3.4.2 Administrative data processing systems

The event sequences in the real-world behaviour model are weakly constrained by corresponding event sequences in the information system behaviour model. These systems will be the most troublesome in deciding how much requirements analysis is necessary, and which of the requirements are mandatory. In some cases real-world events are to be constrained by the information system events. In other cases the information system events are to be recorded without constraining real-world events (although perhaps with a warning). The difficulty is in separating these two cases without undertaking a full ELH analysis, and then finding part of the analysis was unnecessary.

3.4.3 Record Keeping systems

There is a definable real-world behaviour important to the provision of the IS service but the burden of ensuring that the information system represents the current state of affairs is entirely the user's responsibility through ad-hoc updates to the records held in the system. This is the easiest type of system for which to build a package. There is no need to try and develop a generic-behaviour model which is the expensive part of the Conceptual Model. The absence of the behaviour model means that the package can be easily customized through the External Design.

The above system types are more fully described in the ISE Library volume *Accelerated SSADM*. These system types influence the extent of requirements analysis and specification that are necessary. This is examined in Chapter 4.

3.5   Pre-requisites

There are three conditions that must be present if a package implementation approach is to be followed effectively:

- knowledge that suitable package software is potentially available

- acceptance by the users that a less than optimum solution may be implemented

- understanding by the project team of the development approach to be used.

3.5.1   Availability

The availability of an application package should be investigated during the Feasibility Study and again, possibly, during Business System Options. Potential suppliers are sent a formal or informal Request for Information which outlines the scope of the required system and asks for technical information about the availability and suitability of packages.

The difference between the formal and informal Request for Information is that the formal must be issued under EC/GATT regulations, whereas the informal must state explicitly that it is not part of the procurement process (see ISE Library volume *SSADM and Information Systems Procurement* for more information on this topic).

3.5.2   User acceptance

One of the most difficult aspects of using a package is ensuring that the users are fully committed to the philosophy implied by the decision to use a package. In other words, they are prepared to accept that not all their requirements will be fully satisfied. The user's mandatory requirements should be framed in such a way that they must be met if the system is to achieve its business benefits. The users must accept that non-mandatory requirements may not be provided in the ideal way, if at all. At its most extreme they must be prepared to accept, for a package-constrained approach, that the way the package operates may result in major changes to the business processes that the information system is to support.

The reward for compromising the requirements may not just be the provision of a system more quickly or more cheaply. The building of a bespoke system may not be an option. Some organizations have gone so far as to proscribe all bespoke development.

For users who have a choice, the most important benefit is probably reduced risk of failure. Studies have shown that up to 40% of systems are never implemented (either never delivered or rejected) and another 40% have a negative or marginal effect. There is evidence that many systems are abandoned because over-engineering causes spiralling costs and timescales. The limiting of ambition and the reduced timescale for a package solution are significant risk-reduction factors.

| 3.5.3 | Development approach | The big question for project teams is how much of the SSADM analysis to undertake if there is a desire to use a package. The most common short-term motivations to use a package are a need to save development time and development cost. If this is to be achieved the detailed analysis work must be minimized, and the earlier in the lifecycle the decision is made the greater the savings. |

However, the decision to use a package can only be sensibly made with a knowledge of the information system requirements. Obviously some analysis work is required to establish these requirements, but how much? It seems to vary with the type of system under consideration. For example, few people do any recognizable requirements analysis prior to purchasing a word-processing package. On the other hand, the adoption of a Patient Administration System package probably requires as much requirements analysis as the equivalent bespoke system. The reasons why there are these variations are discussed in detail in chapter 4.

3.6 Decision points in lifecycle

The decision to use a package should be made as early as possible to minimize unnecessary analysis and specification work. Note however, that no decision to use a package can be confirmed until it is known whether there is a package in existence that will undertake the required tasks. The first six project-level decision points in the system development lifecycle are summarized below:

- agree scope – Agree the scope of the development, perhaps as a result of a Feasibility Study or by identifying a project in the context of a wider strategy study

- agree business problem situation – Agree the operation of the current system, the problems in the operation of the current system and the functional and non-functional requirements for new system

- Select Business System Option – To solve the business problem, choose an information system solution to some or all of the problems and requirements, at a price acceptable to the customer

- agree Requirements Definition – To solve the business problem, agree in detail the information to be managed by the new system, the business activity patterns to be represented in the information system and the information access facilities to be made available to end-users and other systems

- Select Technical System Option – Select a hardware and software platform for running the new system

- agree Physical Design – Agree the design of computer system which implements the Requirements Specification on the chosen platform.

If there is doubt that a package will be suitable it is preferable to 'posit' a package approach to the development in the knowledge that some additional work may have to be undertaken if the assumption turns out to be wrong. The alternative is to wait until Select Technical System Option before deciding to use a package, with the probable result that a significant part of the requirements definition activity will prove to be nugatory.

The relevant points for making or confirming a decision to use a package are Agree Scope; Select Business System Option; Select Technical System Option.

The relevant point for agreeing the package customization, if any, is Select Technical System Option.

**3.7   Evaluation and
       selection**

This volume is not about evaluating and selecting from competing packages (see Figure 2) or about procurement.

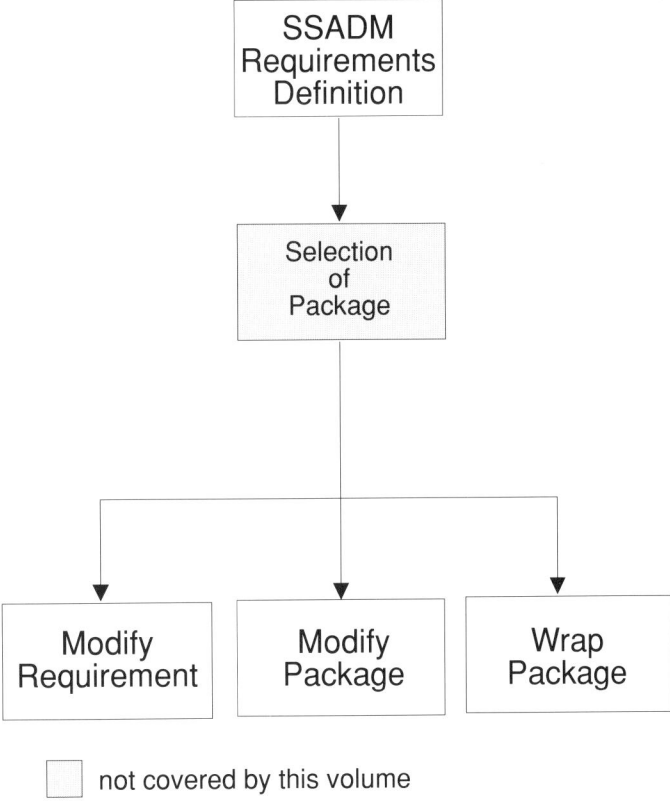

Figure 2: Scope of this volume

The process of evaluating and selecting packages is described in the Appraisal and Evaluation (A&E) Library volume *Overview & Procedures*. This A&E Library volume describes a general procedure for the evaluation of products (including application packages) and specialized versions of the procedure for four kinds of evaluation:

- Business-based Evaluation – carried out in support of an IS strategy to select a software environment for a class of application. The consideration of packages at a strategic level is outside the scope of the present volume

- Evaluation during a Feasibility Study – a preliminary assessment of the requirements for a single application against available products. This is equivalent to the 'Agree Scope' decision point above

- Evaluation for a Project – assessment of products against a completed Requirements Specification for an application. This is equivalent to the 'Select Technical System Option' decision point above

- Product Appraisal – project independent appraisal of a single product, rather than a more detailed evaluation. The general appraisal of packages is outside the scope of the present volume.

Readers wanting a more detailed explanation of the procurement process should refer to:

- *A Guide to Procurement within the Total Acquisition Process*

- ISE Library volume *SSADM and Information Systems Procurement.*

4 Requirements Definition and changes to SSADM techniques

Chapter 2 summarizes the main arguments for and against using an application package. The benefits can be compelling, and the use of packages continues to grow for certain types of system.

Once a decision to try to use an application package has been made, the next question is, how much SSADM requirements definition to undertake? There is a conflict between a desire to specify requirements fully so that the 'best fit' package can be procured, and the desire to maximize the benefits of a 'package-constrained' approach by eliminating requirements definition altogether.

If the decision is taken to eliminate requirements definition it must be realized how high risk a decision this is. If the users' requirements are not defined it is likely that the package selected will not be appropriate and what appears at the outset to save money could become very expensive over the lifecycle of the system.

There are seven factors that influence the extent of requirements definition:

- application type

- user flexibility

- impact of the Conceptual Model

- application maturity

- level of integration required

- risk

- scale.

4.1 Application type

It can be argued that for the procurement of certain types of package, any analysis of the user requirements is not only unnecessary but positively harmful. Would

27

anyone consider even doing an SSADM Feasibility Study to identify the requirements for word-processor software or a spreadsheet. The vendors of these products understand the requirements for this kind of software far better than any individual user organization.

On the other hand, it is equally improbable that an organization would purchase a sophisticated investment management system costing several hundred thousand pounds without detailed requirements definition, probably to a level of detail equivalent to an SSADM Requirements Specification.

There is no universal rule about the level of detail of requirements definition prior to the procurement of the package. However, some guidelines for different classes of system can be proposed.

| 4.1.1 | Infrastructure software | Examples of infrastructure software are word-processors and spreadsheets. There is no software or database to be designed, and therefore no need for an analysis of data or functional requirements. The use of SSADM is inappropriate. |

4.1.1 Infrastructure software

Examples of infrastructure software are word-processors and spreadsheets. There is no software or database to be designed, and therefore no need for an analysis of data or functional requirements. The use of SSADM is inappropriate.

4.1.2 Universal business packages

General ledger; personnel; payroll; purchase order are all examples of universal business packages.

Low cost personal-computer versions of such packages are procured on a comparison of available solutions. These are outside the scope of SSADM.

For more expensive products that justify the use of SSADM, if it is recognized that the required information system is to operate in the 'industry standard' way (ie package-constrained) then an early procurement may be initiated at the end of the Feasibility Study. If it is not certain that all the requirements are supported by available packages, or if a detailed comparison of packages is required, then the production of some of the Requirements Specification (Stage 3) products is needed as a means of assessing how closely the packages fit the requirement.

4.1.3 Specialized business packages

Examples of specialized business packages are vehicle fleet management; investment management and patient administration systems. Some of the Requirements Specification products are required as a means of assessing the packages. A full Requirements Specification is only likely to be required if the package is to be extensively tailored, including modifying the behaviour model.

| 4.1.4 | Modular packages intended to be extensively customized | Banking services and insurance/underwriting are examples of this type of package. These packages use many parameters, and the extent of the tailoring is such that the SSADM activities are not significantly different from those for a bespoke development. |

4.2 User flexibility

If a package is to be used for an application, users must be prepared to accept the limitations on their 'wish-lists' imposed by both package-based and package-constrained approaches.

The degree of requirements flexibility that the users can accept will influence the extent of requirements definition. Lack of flexibility should not be regarded as simply intransigence or the 'not-invented-here' syndrome. There may be genuine reasons why the information system must operate in particular ways:

- there may be organizational 'standing orders' or legislative requirements which are outside the scope of the user department to change

- the absence of knowledge and experience of the application amongst user staff may mean that a highly automated (and therefore probably heavily tailored) system, rather than a record-keeping system, is required

- there may be business critical differences in the way the organization operates, and obtains a competitive advantage, that must be implemented in the information system

- the culture of the organization may restrict the ability to change operating procedures.

The degree of flexibility available may make the difference between package-constrained and package-based, or package-based and bespoke development.

4.3   Impact of the Conceptual Model

The 3-schema specification architecture provides an alternative view of the extent of requirements definition.

4.3.1   Systems without a Conceptual Model

Systems with no discernible Conceptual Model are not really within the scope of SSADM. Infrastructure software packages, process control systems and message switching systems generally fall within this category.

4.3.2   Systems with a conceptual data model

Systems with a conceptual data model but no significant behaviour model are the simplest to implement using an application package. Record keeping systems and some simple administrative data processing systems are of this type.

The exclusion of a behaviour model makes it less difficult to achieve a good fit with the package. For example, the data model for a vehicle fleet management system may be a good fit with the package data model because most organizations hold similar data about their vehicles. However, the processing of the data concerning vehicle use, servicing and inspections is likely to vary between organizations.

Applications that do not have to have a behaviour model implemented in the software are more likely to be able to follow a package-constrained approach. If the match between the required data model and the available package data models is weak, an alternative approach to packages is to use an application generator to implement the data model quickly, with few behaviour constraints and an unsophisticated External Design.

4.3.3   Systems with a conceptual behaviour model

Systems that require a close fit with a full data and behaviour model are less likely to be able to use the package-constrained approach than those without a behaviour model. Real-time systems and some administrative data processing systems fall into this category.

The implementation of a specific behaviour model increases the cost of bespoke solutions, including those utilizing an application generator. Packages that include complex behaviour models tend to be designed for vertical markets, and have a higher cost than universal business packages, reflecting the higher development costs and lower potential sales volume.

| 4.4 | **Application maturity** | The maturity of the application domain influences the level of confidence in a package approach. |

Packages aimed at mature business requirements (for example, payroll; accounts; automatic teller machines) will have, over time, migrated towards a common core functionality. An 'industry standard', or at least clear market leaders, may have emerged, and these packages are less likely to have deficiencies that have a significant impact on the business.

Packages developed for more novel applications (for example, council tax; crime incident systems) have not had time to mature. There is an increased risk of deficiencies, and more requirements definition work is likely to be necessary to ensure that all the essential requirements are satisfied.

| 4.5 | **Level of integration required** | Increasingly packages are being procured as one or more elements of a set of integrated applications. The need to establish the ability of an individual package to be integrated with other applications will influence the requirements definition activities. |

A package-constrained approach can still be used, but there is an additional risk that the package can not be interfaced satisfactorily with existing or planned applications. Some analysis of the interface requirements (ie including an analysis of alien systems) will reduce that risk.

| 4.6 | **Risk** | Reduction of the risk of failure may be one of the reasons for selecting a package approach in the first place. The risk to the business of an inadequate information system also influences the level of requirements definition. |

Purchasing, say, a personnel information system that proves not to support some of the desired data or events is unlikely to have a major impact on company profits (unless the company is a recruitment agency!). In this case procuring a proven package based only on non-functional requirements may well be most effective.

The same approach is unlikely to be acceptable for, say, an insurance company purchasing an underwriting package. The absence of support for certain functional requirements may have a major impact on the business. The risk can be reduced by identifying those requirements and ensuring they are supported by the chosen software.

4.7   Scale

The scale of the expenditure on the package software will influence the rigour of the evaluation process, and therefore the extent of requirements definition. Even infrastructure software, whose procurement usually involves little or no formal requirements definition, may be the subject of detailed analysis of requirements if several hundred licences are being purchased.

However, this analysis often consists of comparing packages against each other, rather than against an independently defined requirement.

4.8   Requirements Definition philosophy

4.8.1   Solution-driven or needs-driven

At the low-cost end of the packages market, purchase decisions are almost always based on an examination of the available solutions (eg PC magazine reviews of competing spreadsheets or accounting software). There is little doubt that for certain classes of system this solution-driven approach is the most effective.

There are those who argue that the solution-driven approach should be extended to other classes of system, where some formal requirements definition is necessary. It has been suggested that system requirements should be reverse-engineered from an examination of the data and functions in available package software. It is claimed that an agreed set of requirements can be more quickly established by this method, and that it avoids the problem caused by specifying requirements unconstrained by available solutions (ie there is no solution). Apart from the philosophical and procurement issues there are significant practical difficulties with this approach.

Firstly, reverse-engineering a package to establish the requirements is not as easy as it sounds. The internal data structures in some packages contain coded entity (record) types, and do not identify the business entity type they represent. There may also be provision in the data structures for user definable entities and attributes. How are these converted to business entity types?

Assuming a data model can be constructed from an examination of the package, it represents a normalized view of the data in the package. It does not represent the business entities and business rules (relationships), ie the problem description. This is critical. The scope of a system is determined by firstly agreeing the set of business activities to be supported, and then defining the extent of the information system support.

The problem description will contain requirements that can be potentially satisfied in more than one way. Some solutions will rely on explicit support within the boundary of the IT system, others will require procedures outside the boundary of the IT system. A requirements model developed from available packages is a model of only the IT part of the system. There is no agreed model of the business activities to be supported. It is akin to deriving the required system model entirely from the current system model.

Finally, a system model derived from the packages represents only the data and processes available in current versions of the software. Over the life of the information system the scope and functionality of the chosen package may change. Later versions of the package may provide facilities to meet requirements that could not be met in the first implementation. A needs-driven requirements model which defines the total information system requirement must therefore be maintained, against which the provision of new solutions can be evaluated.

SSADM is mostly a needs-driven method, and this philosophy is maintained in this volume. Solution-driven approaches are successful for infrastructure type packages and for low cost PC-based packages, but these are generally outside the scope of SSADM. However, there are situations within an SSADM project when part of the system can be provided by embedding a package chosen using the solution-driven approach.

4.8.2 How much
        Requirements
        Definition?

The seven factors described in sections 4.1 to 4.7 are concerned with the extent of requirements definition. Having established that some needs-driven requirements analysis is essential, projects need to determine how much requirements definition is necessary. The requirements definition activities can be limited in two ways:

- reducing the scope

- producing less detail by considering mandatory requirements and not desirable ones.

The scope of requirements definition is reduced by comparing the scope of the business activities to be supported with the scope of the information system support provided by available software.

There is no universal solution to the need to define requirements in less detail. In many cases the production of a detailed Requirements Specification in the SSADM style is not appropriate. Instead a prioritized list of data and processing requirements (possibly supported by a data model and function catalogue) is produced. The introduction of a solution bias must be guarded against, as this could undermine the usefulness of the evaluation of the proposed solutions.

The prioritized list of data and processing requirements is similar in content and format to the SSADM Requirements Catalogue, and this term has been adopted in subsequent sections. In the case of the more expensive modular packages that use many parameters, the level of tailoring possible may mean that the requirements definition is scarcely less detailed than for a bespoke development.

The guidance that follows encourages the procurement of a package as early as possible in the development lifecycle, against a specification containing only sufficient detail to allow an effective evaluation of competing products but giving a clear definition of business needs. The approach recognizes that, for package-based systems, further detailed specification work will have to be undertaken after a package has been procured, but only for those elements of the information system not supported by the 'core' package.

4.8.3 Limiting requirements

The most cost-effective approach for matching information system requirements to a selected package is likely to be via limiting the requirements, rather than via extending the package.

If the ideal is taken to be minimization of the software development effort, the following list summarizes possible approaches in order of preference:

- package-constrained – Change the method of working to eliminate the need for requirements not supported by the package

- package-based with package extensions – Changes to the package are restricted to those that can be achieved using the package extensibility features provided by the vendor, eg generate extra reports using the report-writer, providing an ad-hoc query language

- package-based with bespoke extensions – The core package is not changed, but bespoke software is used to extend it, eg translation software to interface the package to other systems

- package-based with modifications to the External Design – The core package is modified, but the changes are limited to the External Design, eg changing the layout of screens; manipulating input and output data

- package-based with modifications to the Conceptual Model – This approach is to be avoided if at all possible. Changes to the Conceptual Model are complex and have a significant adverse impact on analysis and design costs and maintainability.

The above are not of course mutually exclusive, but can be used as a guide when deciding whether individual requirements are to be supported in the computer-based information system, and how. There will often be more than one way of meeting a specific requirement.

In general there are often substantial cost savings to be made by following a package's standard approach and changing the organization's current method of working, rather than modifying the package to conform to current methods. However, there are many reasons, often political and cultural rather than technical, why modifying the requirements may be unacceptable.

4.9 Feasibility Study

An option of using an application package should always be considered when conducting a Feasibility Study.

Many projects which have an indication at project initiation time that a package solution is likely to be chosen, omit a Feasibility Study. However, a Feasibility Study has even greater importance for projects planning to use a package. It is crucial to identify as early as possible whether a package option is available. A procurement may be initiated for some systems at the end of the Feasibility Study. Omitting the Feasibility Study will result in the procurement being deferred until after Business System Options, which may be unnecessary in some package-constrained cases.

In addition to its usual objectives the Feasibility Study must:

- establish with users that a less than optimum solution to the business problems may be implemented (section 4.2)

- identify that potential packages for the application exist

- decide on the development approach to be used

- define the procurement strategy.

4.9.1 Identifying potential packages	As part of the Feasibility Study the project must investigate the availability of software packages to meet some or all of the requirements.

Initially this is likely to be done informally, drawing on the existing knowledge of the members of the project team and referencing software catalogues, to establish if packages exist for the application domain.

If relevant packages do exist a Request for Information may be sent to suppliers, which outlines the scope of the required system and asks for technical information about the availability and suitability of the packages. If the Request for Information is an informal one (see ISE Library volume *SSADM and Information Systems Procurement*) subsequent procurement is not restricted to packages identified at this point. The information received from suppliers may provide a useful source of ideas for defining and agreeing the scope of the system with the users. |
| 4.9.2 Development approach | By the end of the Feasibility Study the development options should have been considered, these are: |

- package-constrained

- package-based

- bespoke.

At this point in the lifecycle insufficient information is usually available to determine which of the package-based approaches is appropriate.

If the initial investigation of packages indicates a strong fit with available package software, and there is a high degree of confidence that a package-constrained approach is appropriate, an early procurement may be initiated without further analysis of the functional requirements. Low cost packages that represent some form of 'industry standard' are most likely to be procured in this way.

If the fit against available packages is shown to be weak the project may proceed as a bespoke development from this point.

If the package fit is between the above two extremes (probably the majority of cases) further investigation of the requirements will be undertaken in the Requirements Analysis Module. This does not mean that a package-constrained approach has been eliminated, simply that there is insufficient certainty about the suitability of the packages, perhaps where there is no clear industry standard.

However, it is important that the Feasibility Study resolves the bespoke versus package argument if at all possible. Although the decision can obviously be revisited at subsequent decision points, if a planned bespoke project reverts to a package project, some of the requirements definition effort may be nugatory. If the change is from package to bespoke, analysis work may have to be repeated with a user-driven requirements focus.

4.9.3   Procurement

If the conclusion of the Feasibility Study is that a standard package can be procured to meet all the essential requirements, a procurement can be initiated immediately after the Feasibility Study.

The Operational Requirement produced is less detailed than for later procurements. The main document for expressing the requirements is the Requirements Catalogue. There may be functional requirements derived from the Feasibility Study, but these are likely to be very general. To an extent the adequacy of the functional requirements is being taken on trust in a procurement this early in the lifecycle. The evaluation criteria for selecting a package are mainly non-functional requirements, for example size of customer base; support services; vendor stability.

Procuring a package at this point relies on:

- the package being low cost

- confidence in the package-constrained fit being high

- the risk to the business of failure being low.

4.10   Full Study

A Feasibility Study or an equivalent activity as a preliminary to the Full Study is essential. The scope of the project is reviewed at the end of the Feasibility Study, and may be modified in the light of knowledge gained from the responses to the Request for Information. The scope of the business activities covered by the system may be reduced, thereby reducing the scope of the requirements definition in the Full Study, and offering an improved chance that a suitable package will be found.

4.10.1 Requirements Analysis

Stage 1 in SSADM is concerned with modelling the 'current environment'. Since the business activities have been constrained (in the Feasibility Study) by what is available in packages, the information system requirements to support the business activities are also constrained.

With the above exception Stage 1 is conducted in a similar manner to default SSADM. However, the knowledge of typical package facilities gained from the Request for Information can be used to help define the requirements in those business areas where there is no equivalent 'current system' available.

As with default SSADM, the Stage 1 analysis should not be conducted in exhaustive detail. An objective is still to reach the next decision point (Business System Options) as quickly as possible. Specifying requirements in too much detail may be wasted effort.

However, sufficient detail of the requirements must exist at the end of the Stage to allow the costs of all the alternative development approaches (including fully bespoke) to be evaluated. It is not the analyst's responsibility to suppress user requirements to try and force a package solution prior to Business System

Options. There may be resistance to accepting a package solution. Enough analysis must have been done to show the cost of the alternative.

4.10.2 Business System Options

When the requirements have been defined they are prioritized. Some requirements will be mandatory – if they are not supported in the chosen package the package will have to be modified or extended. Others will be capable of being satisfied by alternative procedures outside the software, and therefore will be classified as only desirable requirements for the procurement. The requirements should be reviewed with user management to confirm they are still valid. Most attention is paid to requirements not fully supported by available packages, as they have the greatest impact on the development approach, and hence on the cost of the system. In practice some requirements will be supported in only a subset of the packages. In effect the requirements are compared against a kind of informal generic model of expected package facilities derived from the packages examined so far.

The solution to each requirement should be classified as:

- supported within a generic-package model (including via the use of vendor provided facilities allowing for the use of parameters)

- supported by the development of a set of manual procedures

- requires the development of separate bespoke code extensions

- requires modification of the package

- requires changes in working practices so that the requirement is no longer necessary.

Some requirements will inevitably be incorrectly classified at this point. This is unimportant as they will be re-examined once a specific package has been selected or during the procurement process.

The definition and selection of Business System Options is procedurally the same as for default process description given in the SSADM Reference Manual. The above categories are used to group requirements to define a minimum (package-constrained) option, and a range of package-based options.

4.10.3 Development approach

In addition to fixing the system boundary a decision must be made about the development approach to be used. The package fit considered in the Feasibility Study is re-evaluated here. Further information about the previously identified packages may need to be solicited.

If the chosen option is to use a package-constrained approach then a procurement may be initiated without further specification of the functional requirements. The risks of this approach however have to be borne in mind, as work which would have been done later might have affected the choice of package. Only a very limited set of requirements will have been identified at this stage, so an early decision to procure needs careful consideration with the risks of making the wrong decision evaluated.

If none of the package solutions can provide an acceptable system, then the project may proceed as a bespoke development from this point.

If a package-based option is chosen, further specification of the requirements will be undertaken in the Requirements Specification Module.

Choosing the development approach is not just a function of how well the available packages fit the requirement. Other factors to be considered include:

- the availability of staff to develop bespoke software (including package extensions)

- the availability of the package source code to enable it to be modified

- the need to interface the package to other systems

- implementation timescales

- system performance.

4.10.4 Requirements Specification

The product of a Full Study when a package implementation is intended is not necessarily a full Requirements Specification document as defined in SSADM. For the majority of systems a prioritized list of data and processing requirements, supported by a data model and function definitions and non-functional requirements, will be sufficient.

The level of requirements definition must only be sufficiently detailed to facilitate the final selection of a package against meaningful and reasonably specific business needs. Since there is usually more than one way of meeting a specific requirement, it is preferable to defer a decision on the solution until after a package has been selected. The features of the package software can then be exploited to provide the most effective solution.

If a package-based option is chosen the Analysis of Requirements produced in the Requirements Analysis Module is updated. Requirements outside the system boundary are deleted, and the approved 'new requirements' are added to the Logical Data Model and Data Flow Model in the same way as for default SSADM (Step 310 and 320). A Statement of Requirements may need to be issued for complex systems that include a package element (see *SSADM and Information Systems Procurement*).

How much of the remaining SSADM requirements definition work should be undertaken prior to a procurement depends on project circumstances, particularly three factors discussed earlier:

- extent of the Conceptual Model

- level of integration required

- risk to the business.

The minimum additional requirements definition work for a package-based project is the production of Function

Definitions and Input/Output Structures. The definition of system inputs and outputs provides a more detailed and objective method of evaluating the External Design of packages than using the Data Flow Model.

If the system under consideration is of a type that includes a significant behaviour model, then the production of Entity Life Histories will also be necessary.

The package evaluation must assess how well the behaviour model is supported by the different products.

If a behaviour model is not part of the requirement, Entity Life Histories can be omitted. If only a small number of behaviour constraints need to be included (eg, a business rule that says a reservation event must precede the passenger check-in event) they should be expressed in the Requirements Catalogue. The major business rules are likely to be included in vertical market packages, or there may be other ways of implementing the constraint without undertaking a full behaviour analysis and design. Of course the problem remains of how the behaviour rules are identified without life history analysis.

Although entity life history analysis may be necessary in a limited number of cases, Effect Correspondence Diagrams will rarely be necessary at this point. They may need to be produced later for those events that are outside the scope of the chosen package, but are not necessary for assessing the package. Similarly, Enquiry Access Paths are not generally necessary. The packages can be assessed against the enquiry Function Definitions and associated I/O Structures for enquiries.

If a package is to form part of a larger integrated system, or be linked to other systems, particular attention needs to be paid to understanding and defining the interfaces.

Outweighing all the other influences on the extent of requirements definition is the risk factor. The greater the degree of uncertainty concerning the suitability of package software to fulfil the users' requirements, the more specific the description of the requirements needs to be.

4.10.5 Procurement

This section is a high-level overview of the procurement process (see Annex A for Product Breakdown Structures of SSADM products available at various points for use in procurement). If a procurement is to be undertaken the ISE Library volume *SSADM and Information Systems Procurement* should be referred to.

If the Business System Option is package-constrained a procurement can be initiated without further requirements definition.

All the SSADM products will need to be updated to reflect the chosen Business System Option.

As with an early procurement after the Feasibility Study, the absence of a detailed definition of the functional requirements means that the adequacy of the package is to some extent being taken on trust. If there are uncertainties about the package's suitability, a more detailed specification that can be used to eliminate the unsuitable ones is required. Procurement must be delayed until after the Requirements Specification Module.

Procurement for package-based projects is undertaken after the Requirements Specification Module. SSADM Stage 4 is largely supplanted by procurement activities. The minimum set of SSADM products included in the Operational Requirement are:

- Requirements Catalogue

- Logical Data Model

- Data Catalogue.

SSADM products that may also be provided are:

- Data Flow Model

- Function Descriptions which include the Input/Output Structures

- User Roles.

Entity Life Histories are only included if the package's ability to support a particular behaviour model forms part of the evaluation criteria.

4.11   Techniques

Changes in the role of the SSADM techniques in the Feasibility Study and Full Study are summarized below.

4.11.1 Logical Data
        Modelling

The data model produced in the Feasibility Study represents the business entities and rules. This is constrained to a model supported by available packages, prior to developing a data model representing the data entities and relationships.

The Logical Data Model is the most effective product for comparing individual packages against the user requirements. If the vendor can supply a data model of the application package, establishing the fit against the project data model is straightforward. However, data models of packages are not always available.

4.11.2 Requirements
        Definition

The approach to requirements definition is not radically different from default SSADM, but the level of detail will vary depending on the application type and on the assessment of risk.

4.11.3 Relational data
        analysis

Relational data analysis is not used in either the package-constrained or package-based developments, except informally to help construct valid data models.

4.11.4 Function Definition

The Function Definition technique is broadly unchanged from default SSADM. Function Definitions are usually developed only for package-based projects.

4.11.5 Entity-Event
        Modelling

Entity-event modelling is only used for package-based projects where support for a particular behaviour model is part of the package evaluation criteria.

4.11.6 Specification
        Prototyping

Specification Prototyping is not used in either the package-constrained or package-based developments.

**4.12   SSADM products
        for evaluation**

The product set in the Operational Requirement used for evaluating packages for package-based projects, and for package-constrained projects that have functional requirements defined in more detail than in the Feasibility Study, consists of:

- Requirements Catalogue

- Required System Logical Data Model

- Required System Data Flow Model

- Function Definitions

- Input/Output Structures.

The vendor responses can therefore be evaluated against an independent evaluation model, rather than comparing packages directly with each other.

The risk of this approach is that the additional requirements are sufficient to ensure that none of the available packages are satisfactory for package-constrained projects, or the extent of tailoring for package-based projects becomes equivalent to a bespoke development. In the case of the former the risk can be avoided by classifying, if possible, the additional requirements as 'desirable' rather than 'mandatory'.

Avoiding the latter situation relies on the requirements limitation approaches described earlier.

5 Completing the design

This chapter describes the completion of the design for those projects that require modifications and extensions to the basic package, or development of bespoke software to interface to the package.

In the case of package-constrained development, neither SSADM Stage 5 (Logical Design) nor Stage 6 (Physical Design) is relevant.

In the case of package-based development, Stages 5 and 6 need only consider the extensions and modifications to the package.

The true extent of the additional software requirements may only be certain after a specific package has been selected. If the additional software is extensive it may be necessary to re-scope the project and undertake additional requirements definition work to define the residual system in more detail.

5.1 Rescoping the project

At the end of the Requirements Specification Module, or possibly as a result of unsatisfactory responses to a package procurement, it may be recognized that the additional software required is so extensive that a package-based solution is not appropriate. 'The devil is in the detail,' and although there may have appeared to be a reasonable fit at the level of say Data Flow Models, more detailed analysis may have revealed large numbers of differences at event and data item level.

The package-based approach may be abandoned at this point in favour of fully-bespoke. In this case the project will need to go back to the Business System Option decision point and re-examine the options. Requirements may have been suppressed in the chosen option to ensure a package solution was viable, typically motivated by the desire to shorten timescales. If a bespoke development is now the only option, requirements previously excluded may be reinstated. Starting with the revised Business System Option, a full Requirements Specification will then need to be produced in the standard SSADM manner.

An alternative to fully-bespoke is to draw the boundary of the package-supported system more tightly. The package is used to provide only a partial solution, with the rest being provided by bespoke software that is decoupled from the package (rather than being seen as an extension of it). For example, a separate database may be extracted from the package database to support a functional area not included within the package. The integrity of the package software is retained, testing and maintenance problems are avoided and the users receive the extra functionality in the way they wish. In this case a more detailed Stage 3 is necessary, but only for those parts of the system outside the scope of the package, plus an analysis of interfaces. Detailed design of the residual software will proceed as for default SSADM. The isolation of some requirements into a separate sub-system may make it possible for the package element to revert to a package-constrained approach.

5.2 **Mapping the Requirements Specification**

Initially the requirements are mapped to proposed solutions during Business System Options. However, the solutions at that point are based on a generic view of available packages. Once a specific package has been selected the intended solutions need to be reviewed and cross-referenced to the actual solutions. The solutions are classified as:

- supported within the selected package, including those supported via vendor provided customization features

- requires the development of alternative manual procedures

- requires the development of separate bespoke code extensions

- requires modification of the package.

The specification of requirements produced by Stage 3 with its solution cross-references must be maintained. Later versions of the package may provide facilities to meet non-mandatory requirements that could not be met (by the software) in the first implementation. Maintaining the Requirements Specification will allow improved solutions to be identified and implemented, if they become available.

It could be argued that to take advantage of later versions of a package, the Requirements Specification should represent all the user requirements, unconstrained by what is available in packages. Those that could not be implemented at the first implementation, could be included as the functionality of the package changed. In addition to the cost and time implications, the problem with this approach is that the scope of the final implementation is never known. In effect the Business System Options would have to be reconsidered whenever the package software was upgraded.

| 5.3 | **Wrapping the package** |

Wrapping the package is concerned with designing the facilities that are to be implemented in addition to the package. A general philosophy of the package-based approach is that it is preferable to enhance the system functionality by designing and implementing loosely coupled additional components, rather than modifying the package code. These additional components include both extensibility facilities produced by the vendor, and fully bespoke extensions.

The most common extension approaches are summarized below:

- additional reports are defined using a report-generator supplied with a package

- additional enquiries (including ad-hoc requests) are supported by providing a user query language

- bespoke front-end procedures may be developed to load the database, either by converting data from existing systems or file formats, or as a means of providing a 'friendlier' input interface than provided by the package

- the data maintained in the package database is exported to another database that can be used to provide the missing functionality. The extra functionality may itself be provided by a package, for example the data may be exported to a spreadsheet file for further manipulation

- some packages are modified at run-time by the use of parameter files

- translation interfaces are provided to allow the existing systems to communicate with the package

- some packages contain customer-definable data items linked to particular entity types.

5.4   Modifying the package

5.4.1   Impact of changing code

Modifying the package code means that there must be access to the source code, which is not always available. Even if the source code is available, making changes may invalidate the vendor's warranty or other service agreement. Testing the modified software will be the responsibility of the purchaser, and the modifications will have to be applied to future releases of the package which may not necessarily be compatible with the baseline changes.

Having the modifications made by the vendor may mean service agreements are still valid, and shifts the testing responsibility. The vendor may even have libraries of 'off-the-shelf' modifications produced for other customers.

The principal motivation of many organizations adopting packages is not to save development effort, but to avoid maintenance effort. Modifying a package is often proscribed for this reason.

5.4.2   Modifying the External Design

Changes to the External Design are likely to be the most common. Packages often provide facilities for limited modifications to inputs and outputs, and renaming of data items. Some even provide the capability to change the language of the External Design, without changing the Conceptual Model.

The philosophy of extending the package via loosely coupled additional components means that the majority of the extension approaches listed in section 5.3 are concerned with changing the External Design. The package may also be made more apposite by:

- using vendor-supplied screen-painters to perform limited modifications. These are usually restricted to suppressing output fields or changing the position of displayed fields

- using a data item provided for one purpose for another purpose

- renaming data items to resemble more closely local terminology

- using customization parameters provided within the package.

5.4.3 Modifying the Conceptual Model

Changes to the Conceptual Model should be avoided if possible. Changes to the data model are less troublesome than changes to the behaviour model, but should be limited to adding extra data items into existing entity types. Adding entity types and changing relationships will require changes to the behaviour model.

Changes to the behaviour model are particularly difficult to insulate from other parts of the system. If events not included in the base package are to be added it is likely they will be constrained by (or have a constraining effect on) events that are included. In order to ensure the processing includes these constraints it will be necessary to define Entity Life Histories including both the 'package events' and the additional events. The valid event sequences in the package may be defined in such a way that establishing the life histories is extremely difficult. This is certainly a level of design complexity that most users of packages are seeking to avoid by their choice of development approach.

5.4.4 Modifying the Internal Design

As a general rule manipulation of the Internal Design will not be possible unless the source code is purchased.

Annex A: Requirements Specification Product Breakdown Structures

The following diagrams illustrate, in a simplified fashion, some of the SSADM products that are used to form the specification of requirements for the procurement of a package. Three sets of products, representing different levels of detail, are provided. They correspond to the three possible procurement points described in Chapter 4:

- after the Feasibility Study (Figure A1)

- after Business System Options (Figure A2)

- after Definition of Requirements (Figure A3).

For further details of the SSADM products that are used in the procurement process see ISE Library volume *SSADM and Information Systems Procurement*.

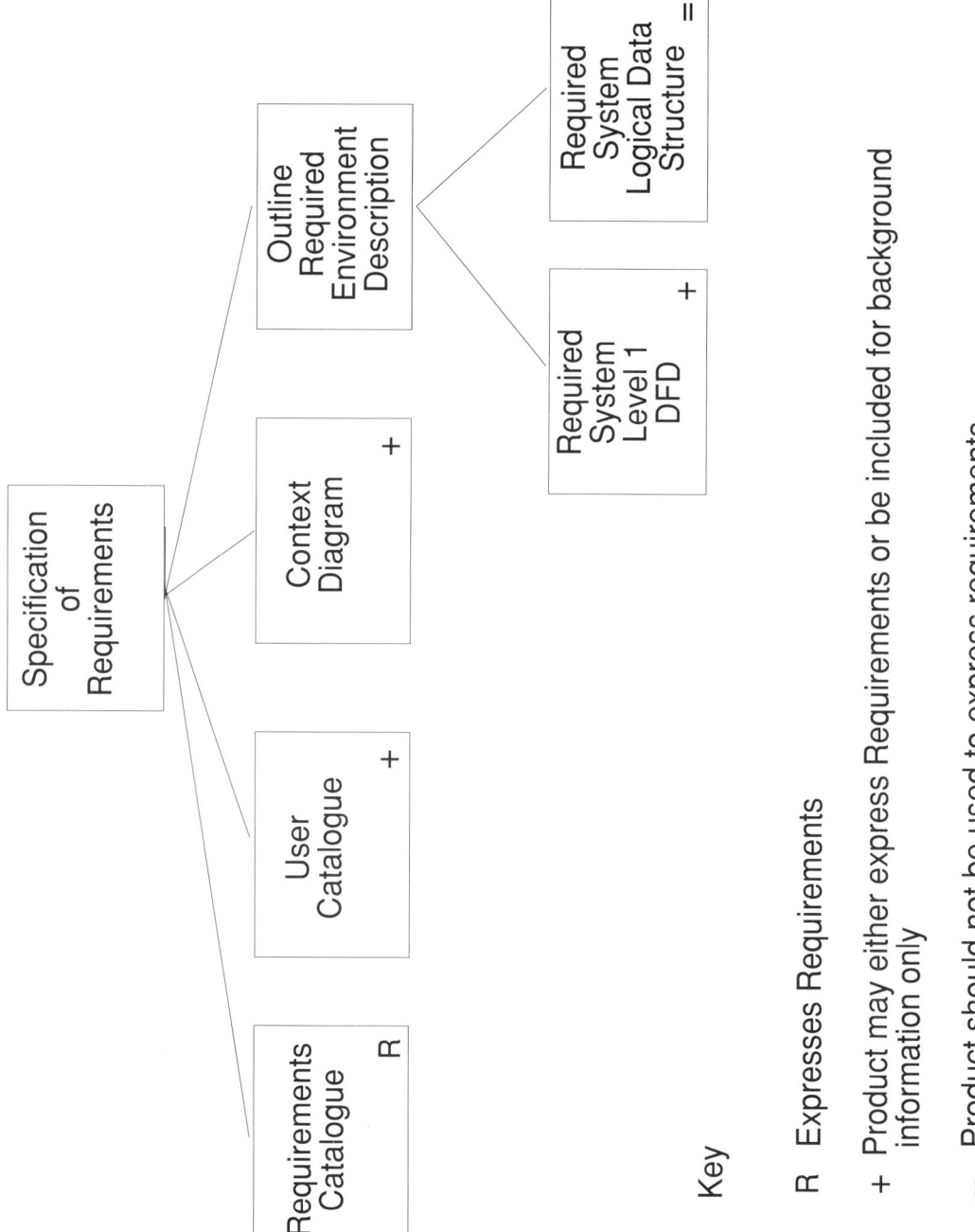

Figure A1: Specification Products for procurement after a Feasibility Study

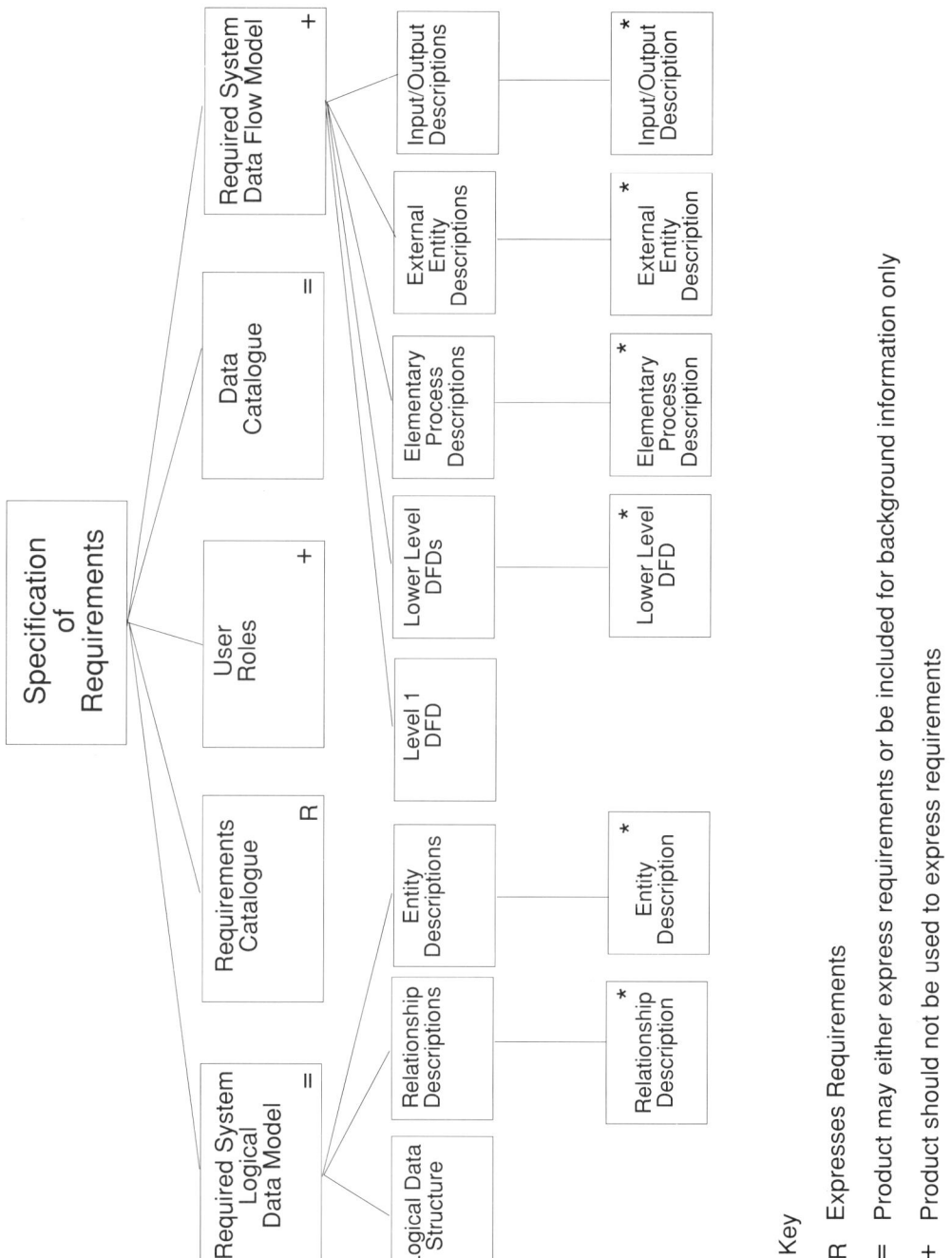

Figure A2: Specification Products for procurement after Business System Options

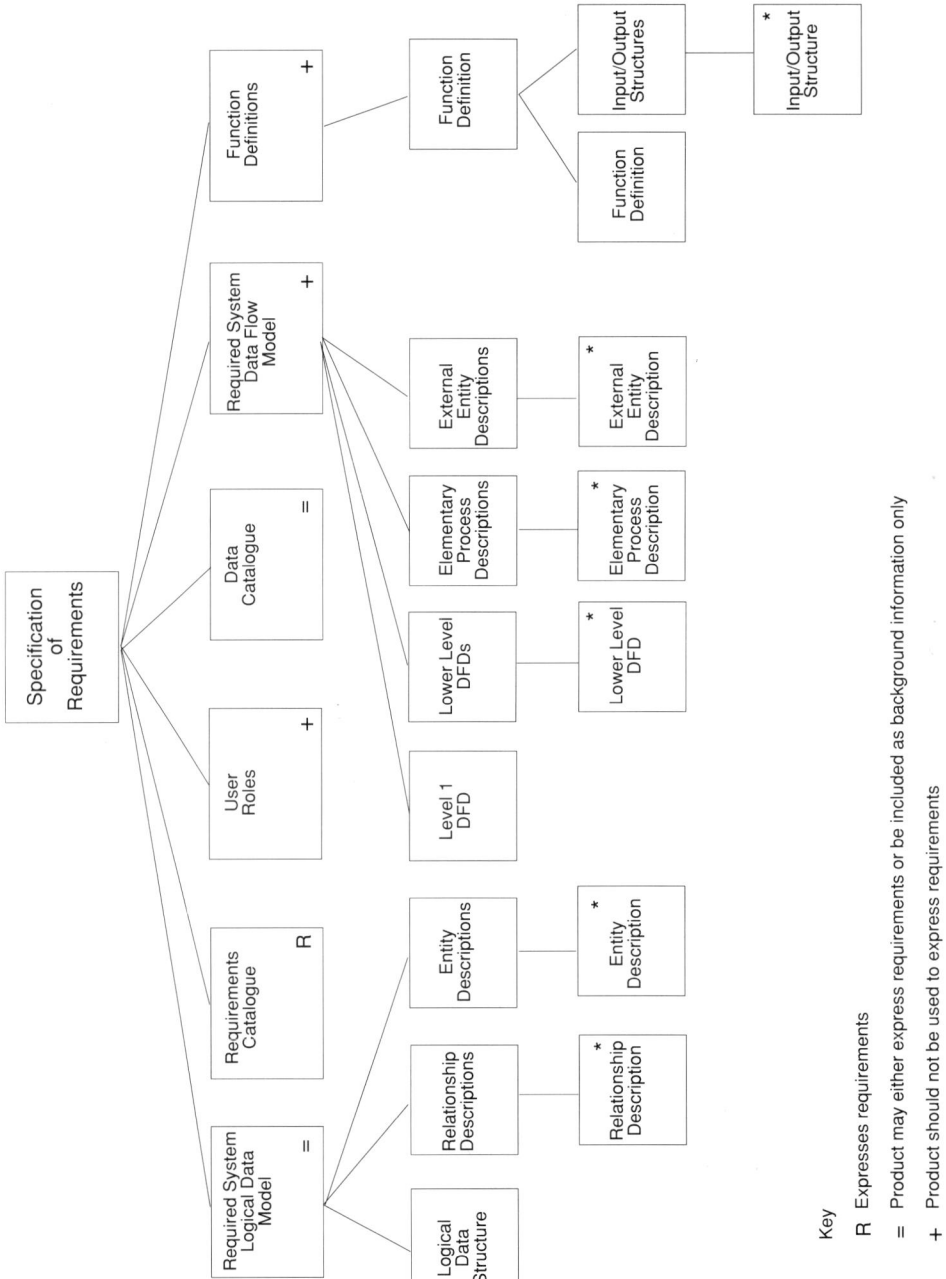

Figure A3: Specification Products for procurement after Definition of Requirements

Annex B: The 3-schema specification architecture

This volume uses the 3-schema specification architecture, as defined below, to define where and when SSADM can be accelerated.

SSADM provides a 'default' Structural Model, containing all the analysis and design activities and products which may be relevant to the development of a large information system. It is a generic approach, capable of being used in a wide variety of project circumstances. SSADM should be tailored to meet the specific requirements of individual projects.

SSADM Version 4 has been designed to allow maximum flexibility. There are many ways SSADM can be adapted for various situations and development environments.

In order to ensure that all such guidance is consistent and compatible with SSADM's underlying principles, the concept of a System Development Template has been introduced as part of the philosophy of the SSADM. This template breaks system development into a number of distinct areas of concern. For the purposes of this guide these major areas are (see also Figure B1):

- Investigation

- Specification

- User Environment

- Decision Structure

- Policies and Procedures

- Construction.

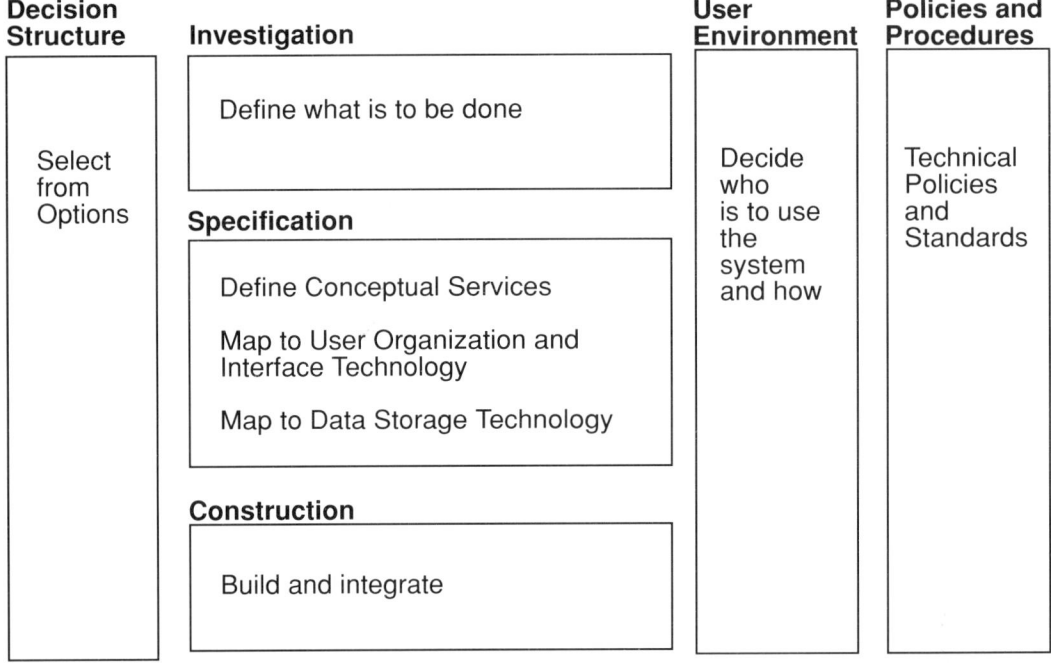

Figure B1: A System Development Template

The 'specification' component of the System Development Template is further subdivided to contain three important classifications of system products:

- Conceptual Model

- External Design

- Internal Design.

These three parts are collectively known as the 3-schema specification architecture and are described in outline within this chapter.

Customized versions of SSADM should be mapped on to the System Development Template with particular attention being paid to the 3-schema specification architecture.

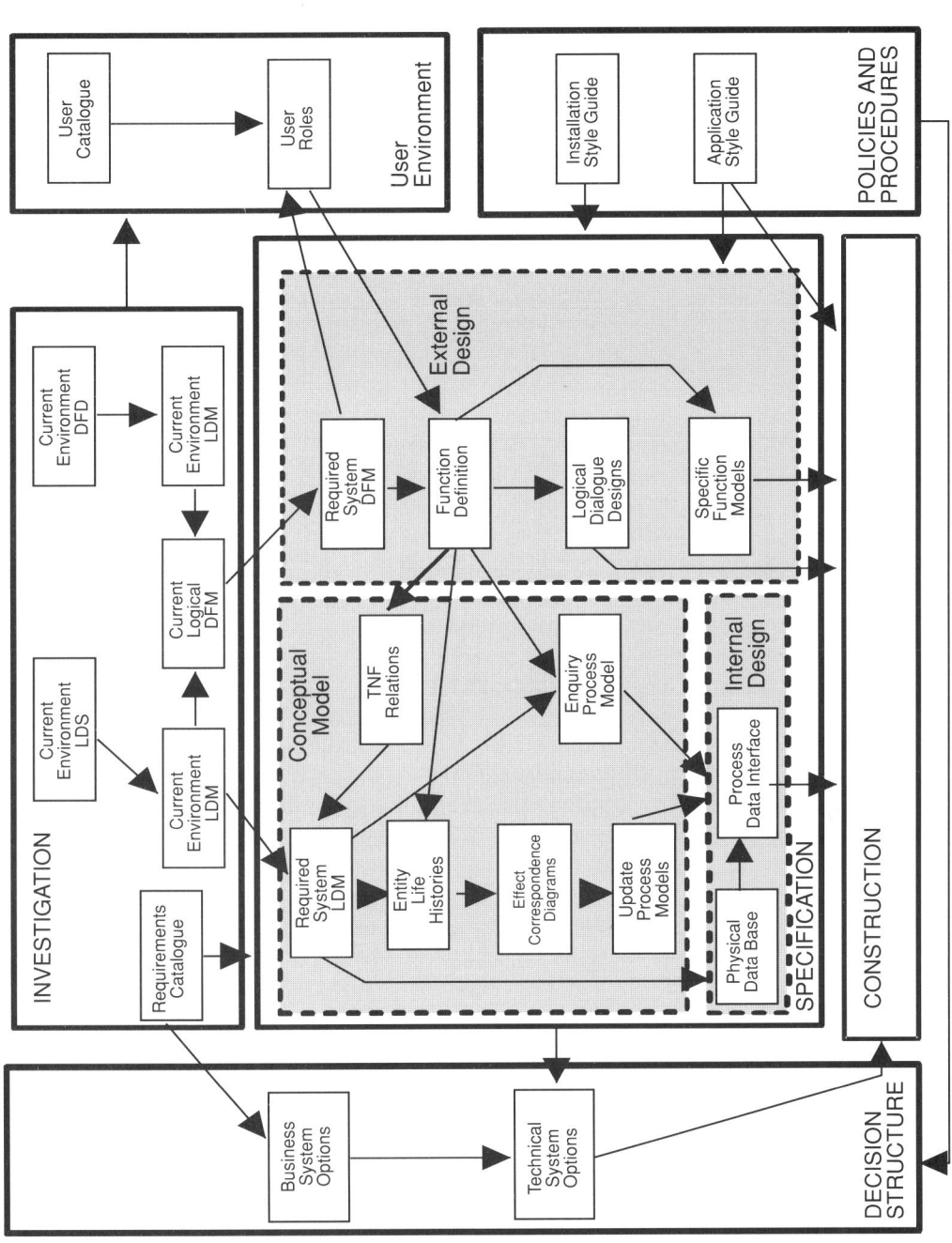

Figure B2: SSADM Version 4 mapped to System Development Template

The System
Development
Template and the
3-schema
specification
architecture

Figure B2 provides an example of the use of the System Development Template. In Figure B2 the products of core SSADM as defined in the Version 4 reference manual are mapped on to the System Development Template.

The 3-schema specification architecture can be used to divide system specification into three parallel development strands, as will now be explained. The three strands extend the separation of concerns already demonstrated within the SSADM Version 4 reference manual via the Universal Function Model.

The Conceptual Model comprises the essential business rules and knowledge, expressed in a Logical Data Model and Entity-Event Model. This is a system model which is independent of the user-interface, and portable between implementation environments. It can be implemented as logical database processes which apparently read and write entities in the Logical Data Model. For the design of the Conceptual Model it is possible to believe that there is in some sense a 'right answer'. Through the use of stereotype components and a disciplined approach to Entity-Event Modelling, the designer can produce a highly objective procedural specification of the database processing. All elements of the Conceptual Model will ultimately be implemented in the computer system.

The External Design comprises the user interface (data definitions for input/output files, screens and reports; and process definitions for dialogue and batch input/output programs). There are two levels of consideration within External Design: the mapping of business services on to User Roles, and the mapping of these Roles on to the interface technology. The External Design depends on trade-offs between a number of different things (organization structure, ergonomics, system efficiency, end-user input/output device technology, arbitrary preferences of particular users, audit principles, security, user politics, etc). So any method for designing the I/O processes must be creative, it must involve inventing solutions. Heuristic approaches such as prototyping clearly have a part to play here.

The Internal Design defines the physical data design (perhaps tuned for performance reasons) and the PDI (elementary data storage and retrieval processes which deal with reading and writing individual records from the physical database, so that conceptual processes can act as though they read and write entities in the Logical Data Model). The Internal Design also depends on trade-offs between a number of things whose relative importance is subjectively defined: time objectives, space objectives, maintainability. Once again, this implies that there is no one 'right answer' and that a heuristic, prototyping approach may be needed.

The 3-schema specification architecture concentrates on those products that will ultimately turn into code. The System Development Template takes a broader view and enables us to identify products which feed into and are developed from those described within the 3-schema specification architecture.

Bibliography

Information Systems Engineering Library

The Information Systems Engineering Library is available from HMSO Publications Centre, PO Box 276, London SW8 5DT.

The following volumes are referenced in this publication:

- Prototyping in an SSADM Environment
  ISBN: 0 11 330582 6

- CASE and the Issues for IS Management
  ISBN: 0 11 330594 X

- SSADM and Information Systems Procurement
  ISBN: 0 11 330627 X

Appraisal and Evaluation Library

The Appraisal and Evaluation Library is available from HMSO Publications Centre, PO Box 276, London SW8 5DT.

The following volumes are referenced in this publication:

- Overview and Procedures
  ISBN: 0 11 330534 6

SSADM Documentation

The SSADM Version 4 Reference Manual is published by NCC Blackwell Ltd and is available from NCC Blackwell Ltd, 108 Cowley Road, Oxford, OX4 1JF.

ISBN: 1 85554 004 5

Information Systems Guides

The Information Systems Guides are available from John Wiley & Sons Ltd, Baffins Lane, Chichester, PO19 1UD. Telephone 0243 779777.

Information Systems Procurement

A Guide to Procurement within the Total Acquisition Process

ISBN: 0 946683 58 1

Glossary

3-schema specification architecture	The 3-schema specification architecture divides system development into three strands of schemas, Conceptual Model, External Design and Internal Design. See Conceptual Model, External Design and Internal Design for information on the individual schemas.
Application Generator	A set of specialized computer programs designed to facilitate the very rapid implementation of computer-based applications, especially on-line applications, with minimal recourse to the use of conventional programming.
Application Software Package	Commercially produced software that comprises a set of computer programs that claim to offer a generalized solution to a defined business problem. The software may or may not be tailorable to met specific requirements.
behaviour model	see conceptual behaviour model.
Business System Options	The set of Business System Options which is compiled so that a selection can be made. Also a Stage of SSADM whose aim is to take the Requirements Catalogue, Current Services Description and User Catalogue and to use this information as the basis on which to decide the most appropriate way for development to meet the business needs.
Conceptual Model	A model defining the essential business rules of the system. It is independent of the user-interface and portable between implementation environments.
conceptual behaviour model	The part of the Conceptual Model that defines the system behaviour rules and knowledge.
conceptual data model	The part of the Conceptual Model that defines the system data and relationships.

Context Diagram	A product that may be drawn to illustrate the initial scope of the proposed system. The diagram concentrates on the major inputs and outputs of the system and shows the external sources and recipients of system data.
Current Services Description	Provides the details of the logicalized current system which, with the Requirements Catalogue and User Catalogue, is output from Stage 1: Investigation of Current Environment.
Data Catalogue	The central repository for all the descriptive information about items of data. This includes physical details which may be found during data flow modelling activities as well as physical design activities. Logical data modelling will provide information about attributes (the logical equivalent to data items).
Data Flow Diagram (DFD)	Shows how services are organized and processing is undertaken. It should be a simple diagram that is readily understood, so that it can act as an effective means of communication between analysts and users.
Data Flow Model (DFM)	A set of Data Flow Diagrams and their associated documentation. The diagrams form a hierarchy with the Data Flow Diagram Level 1 showing the scope of the system and the lower level diagrams expanding the detail as appropriate. Additional documentation provides a description of the processes, input/output data flows and external entities.
Effect Correspondence Diagram (ECD)	Shows all the affects an event has on data within the system and how those effects impact upon each other. Effect Correspondence Diagrams provide access path details for update functions which are used in logical design activities.
Entity Life Histories (ELHs)	Structure diagrams for all entity life histories identified within the system. An Entity Life History is a structure combining all possible 'lives' of every possible occurrence of the entity.

Evolutionary Development   An information system whose initial delivery is unrefined with each further delivery of the system adding more functionality and facilities than were present in the previous delivery.

External Design   A model defining the user-interface aspects of the system, including process definitions for on-line dialogue and batch input/output programs, and data definitions for input/output files, screens and reports.

external entity   Is a source or recipient (or both) of data which exists outside the boundary of the defined system but which communicates with the system. An external entity may be another system, an organization, an individual or a group of people.

Function Definitions   Is the packaging of all details about functions to be included in the Requirement Specification. These details are further expanded during physical design activities. Basic information about a function consists of a Function Definition with one or more I/O Structures.

Incremental Delivery   The delivery of an information system in a series of increments, where each increment delivers an additional part of the External Design or Internal Design. The Conceptual Model is assumed to be stable.

Input/Output Structures (I/O Structures)   Set of documents that records the input to and outputs from a function, or part of a function.

Internal Design   A model comprising the physical database design, and the Process Data Interface (PDI).

Logical Data Model (LDM)   Provides an accurate model of the information requirements of all or part of an organization. This serves as a basis for file and database design, but is independent of any specific implementation technique or product.

Operational Requirement

A document forming a step in the procurement process, containing a complete statement of the procuring Department's requirements, addressed to one or more potential suppliers of equipment or services and designed to draw from each supplier a Full Proposal describing in detail how the supplier could meet the requirements.

Package

See Application Software Package

Package-based

This approach may be used when the major part of the system to be developed is to be implemented using application software packages, but some of the user facilities will be provided by developing bespoke programs, and maintaining data outside the package files.

Package-constrained

A package-constrained development is when the system developed for users is constrained to only those services that can be provided by the application software package with no bespoke work (other than the possible use of a query processor and report generator on the application package files or database).

Physical Design

Is the product that defines the data and processing elements of the implementable system.

Process Data Interface (PDI)

Documents how the Logical Data Model can be mapped on to the Physical Data Design, showing it interfaces with the Physical Processing Specification. It allows the designer to implement the logical update and enquiry processes as physical programs, independently of the physical database structure.

Requirements Definition

The term is used in this volume to include both requirements analysis and requirements specification when the distinction is unnecessary or unhelpful, and to distinguish between the activity and the SSADM 'Requirements Specification' product.

Real Time System

Systems where the real-world events and the information system that supports them run in parallel. An example of this type of system is an airline seat reservation system.

Record Keeping System	This type of system holds information but it is the responsibility of the end-users to ensure that the data is current and correct.
Required System Data Flow Model	see Data Flow Model.
Required System Logical Data Model	see Logical Data Model.
Requirements Catalogue	Is the central repository for information covering all identified requirements, both functional and non-functional. Each entry is textual and describes a required facility or feature of the proposed system.
specification prototyping	Is used to identify and trap errors in the specification of the user requirement and enhance them prior to detailed logical design activities being undertaken.
Statement of Requirement	A document produced at an early stage of the procurement process, which specifies the mandatory requirements of the Department procuring the information system. This document is addressed to suppliers, and they are invited to respond with Mini-Proposals.
Technical System Options	The set of Technical System Options which has been developed so that the system development direction can be chosen. Each option documents the functions to be incorporated and details implementation requirements. Each description is textual with some planning information. Functional elements are taken directly from the Requirements Specification.
User Catalogue	Provides a description of the on-line users of the proposed system. It includes details of job titles and the tasks undertaken by each of the identified users.
User Roles	Is used to document the details for each *user role* identified as having a direct interest in the required system.

Index

Printed in the United Kingdom for HMSO
Dd299872 10/94 C6 G3397 10170